University Cardiff Press

ISBN 0 906449 28 6

British Library Cataloguing in Publication Data
Hoyle, Fred
Quasar controversy resolved.
1. Quasars
I. Title
523 QB860

First published 1981 in Great Britain by:
University College Cardiff Press, P.O. Box 78, Cardiff, CF1 1XL, U.K.

Printed by Qualitex Printing Ltd., Cardiff, U.K.

Contents

For me, the calculation of random probabilities for the spacings within groups of quasars was a defined game as long ago as 1966. Every subsequent discovery of spacings with small random probabilities has therefore hit me with full weight. The accumulated product of these probabilities is by now so exceedingly minute that, even without the Arp–Hazard alignments, I personally can have no doubt at all of the error of the cosmological theory. That the astronomical community should continue to think otherwise is more a matter of sociology than of science. Science is not made by what communities think, but by what the Universe is.

If quasar redshifts are not cosmological, what are they then? Towards the end of the essay I have made a suggestion which has the virtue of answering immediate questions in a satisfactory way. After the essay was written in September 1980, Jayant Narlikar raised the question of the effect of observation on triplet alignments, and the results of independent investigations by the two of us are reported in an appendix.

About 1500 quasars are presently known, many with much detail concerning their individual properties. This treasure-house has been assembled through the efforts of many workers. In a recent quasar catalogue compiled by Adelaide Hewitt and Geoffrey Burbidge as many as 743 references to the literature on quasars are given. Unfortunately there was no opportunity for me to pay tribute to more than a few particular islands set in so great an ocean of information. My narrative involves just two topics, the discovery of quasars and the interpretation of their redshifts. The restriction to this particular story line led, of necessity, to the omission, or partial omission, of many names that have greatly distinguished themselves in the subject. Here I would like to express my debt especially to Margaret Burbidge and Roger Lynds, who have always been generous in supplying me with the latest data, and who somehow had the patience not to become exasperated with my elementary knowledge of atomic spectroscopy. I must mention too the many radioastronomers whose unremitting efforts over the years has led to the discovery of so many of these remarkable objects.

Fred Hoyle

Cumbria
February 1981

But the children of the Kingdom shall be
cast into outer darkness: there shall be
weeping and gnashing of teeth.

Matthew vii, 12

Introduction

Everybody knows nowadays that quasars are compact objects with a star-like appearance when seen directly in a telescope but which have spectra unlike stars. The spectra are those of a quite strongly ionised but diffuse gas, with the emission lines all uniformly shifted to longer wavelengths — redshifted as one says. Although all emission lines are uniformly shifted in any particular quasar, the factor by which wavelengths are increased differs from one quasar to another. The smallest observed redshifts involve an increase of only a few per cent in wavelengths whereas the largest observed redshifts involve about a 350 per cent increase (redshift factor 3.5).

There are three known causes of redshift:

(1) The expansion of the universe,
(2) Local bodies emitting light show a redshift when they move away from us, the so-called Doppler shift,
(3) Light generated in a strong gravitational field becomes redshifted as it emerges into a region of weaker gravity.

Everybody has agreed from the beginning that (3) is not a possible explanation for quasars. The spectrum lines, especially when they occur in absorption, are too narrow to permit strong gravitation, variations of which tend to broaden the lines unacceptably. Moreover the volumes through which gravity could be sufficiently strong are much too small — a *diffuse* gas within small volumes could not produce remotely enough light to account for the emission lines.

Nearly everybody has also dismissed (2). If there were compact objects in intergalactic space, moving at high speeds close to that of light (such as would be required to explain the observed redshifts), surely some objects would move towards us, and for these the spectrum lines would be shifted to the blue — the wavelengths would be reduced? Blue-shifts are never observed, however, and so (2) is also wrong it has been argued.

This leaves only (1), the expansion of the universe, unless we have recourse to some further process so far unknown to science. The view of the majority of astronomers is that the assumption of all manner of fortuitous coincidences is to be preferred to any such notion. Whatever it needs to bolster-up (1) must be correct. This is the conformist position.

What I am going to show in this essay is that the conformist position is wrong, not just probably wrong, but devastatingly wrong. The cause of the redshifts of quasars is either (2), the Doppler shift, or some at present unknown process. Of these, the simple Doppler shift needs more careful consideration. The blue-shift problem, which I mentioned a moment ago, was never a convincing objection. As long ago as 1964 J. Terrell suggested that quasars had all been expelled as rapidly-moving fragments from an explosion which occurred some time ago at the centre of our own galaxy. Inevitably, they would all be receding away from us in that case, so that no blue-shifts could occur. Terrell's proposal has never received much support, however. There are quite a number of difficulties with it, although none of the objections are entirely decisive. A more likely possibility is that quasars emit their light in a beam directed backwards in the manner of the exhaust from a rocket. Inevitably also in that case, only redshifts could be observed, because for those quasars coming directly towards us no light would be emitted forwards in our direction.

Other things being equal, one would prefer to avoid such a highly specific requirement. But there is no intrinsic difficulty in it, no improbability remotely as wild as is needed to support the cosmological possibility (1). Furthermore, we have in another kind of object, the pulsars, an example of an object in which light is definitely known to be emitted as a directed beam. We shall see as we go along that the rocket kind of idea may quite well be the correct theory.

But now let me take you back to what for me was the beginning of the quasar story. It began in the Spring of 1961 with a visit to the radioastronomers at Jodrell Bank. At that time R. Hanbury Brown together with L.R. Allen and H.P. Palmer had just devised the first interferometric method for determining the angular sizes of radiosources. They had found most sources to be about a minute of arc across, but there was a modest fraction that were much smaller — indeed, too small to be measured by the techniques that were then available. As if it were only yesterday, I can hear Hanbury Brown's voice:

> 'We have this family of chaps, less than a second of arc across. We don't know what to make of them.'

The family of chaps were quasars as it eventually turned out.

That visit to Jodrell Bank in 1961 was a major turning point in my thinking. Ten years earlier, in 1951, I had played a minor role in an acid controversy in which the main combatants were Martin Ryle and Tommy Gold. Ryle maintained that radiosources were nearby stars in our own galaxy, whereas Gold and I argued that they were distant galaxies. As it turned out and luckily for us, we were right. By 1961 this was generally agreed. But could it be, I asked myself in 1961, that Hanbury Brown's family of chaps really were Ryle's nearby stars after all? Could it be that, while Gold and I had been mainly right in the controversy of 1951, Ryle had also been partly right? Perhaps because in 1961 I was again in a controversy with Ryle on another issue I was emotionally motivated to reject this possibility. So I jumped instantly to the thought that the Allen–Hanbury Brown–Palmer family of chaps were also extragalactic in their nature. This led immediately to an idea which is common knowledge today, but which was strange and unorthodox in 1961.

From the mid-nineteen-fifties it had been widely accepted in astronomical circles that most radiosources were caused by collisions of galaxies, the diffuse gas in one galaxy impacting the gas in another galaxy. In such a process there was no high concentration of energy. It was rather that the volumes over which the collisions took place were so huge that a low concentration in each unit of volume added up to a very large energy total. What if this picture were entirely wrong I asked myself, what if radiosources owed their origin to the explosion of extremely compact objects at the centres of galaxies? Such objects taken by themselves might be the Allen–Hanbury Brown–Palmer family of chaps.

A few months later I discussed this idea with John Bolton at the California Institute of Technology, finding that he was of a similar opinion. This encouraged me to pursue the matter, which I did in collaboration with Willy Fowler also of Caltech. By mid-1962 we sent our first paper on the subject to the Monthly Notices of the Royal Astronomical Society. This paper, although on the correct lines as it was to turn out nine months later, contained a major error. We attributed the energy of radiosources to nuclear reactions that were somewhat similar to those which take place in ordinary stars. But by the end of 1962, still three months in hand before the actual discovery of quasars, we had seen our mistake. We had seen that the energy had to come, not from nuclear physics, but from gravitation.

Quasars might have been discovered in 1961, two years before they actually were. At just about the time of my visit to Jodrell Bank, Allan Sandage at the Mt. Wilson and Palomar Observatories obtained the optical spectrum of the radiosource 3C 48, a member of Hanbury Brown's family. A short while later, Jesse Greenstein at Caltech also obtained a spectrum of 3C 48, but neither he nor Sandage were able to interpret the jumble of strange lines which they found.

The spectrum was not published. If it happened that you were chatting with Greenstein in his office, and if it chanced that the conversation took a suitable turn, he would take out the slip of glass with the bomb-blast of a great discovery on it and you would be permitted to study the 3C 48 spectrum for a few minutes. As things fell out I never saw the spectrum myself, but John Bolton did. One day I walked with him the quarter of a mile or so which separates the Astronomy Department from the Faculty Club at Caltech. He told me that in his opinion 3C 48 was an extragalactic object, adding in a voice which I can also hear still ringing in my head.

'I think there's a big redshift on the spectrum.'

Bolton had not yet made the study of atomic spectroscopy which he did later, and so was awkwardly placed to interpret the details of the 3C 48 spectrum. My own knowledge of spectroscopy was rudimentary, but with the right idea, that a large redshift was involved, a rudimentary knowledge might have been sufficient. If we had had free access to the spectrum we would undoubtedly have had a shot at solving the line-identification problem. But the spectrum was locked away in Greenstein's office and we would have to explain why we wanted to see it. I think it was simply the fear of seeming stupid in front of experts which prevented us from pursuing the issue. At all events, we were very close, and we too let the opportunity slip.

Even so, John Bolton was to be one of the *dramatis personae* involved in the actual discovery. The others were Cyril Hazard and Maarten Schmidt. The special feature of 3C 48 which had permitted its optical spectrum to be obtained was that this particular member of the Allen–Hanbury Brown–Palmer family happened by good luck to fall at a place on the sky where there were few ordinary stars. Thus although in 1961–62 radio-astronomers could only determine absolute positions on the sky that were subject to considerable uncertainties, the uncertainties did not prevent an optical identification being made in the special case of 3C 48. For other members of Hanbury Brown's family, however, the areas in question of the sky were much more crowded with ordinary stars, so much so that almost everybody despaired of ever being able to distinguish which among so many visible possibilities were the true radio objects. It was Cyril Hazard who spotted a clever trick for overcoming this apparently insuperable difficulty.

Hazard noticed that the Moon in the course of its 18.6 year cycle crossed the positions of one or two members of the Allen–Hanbury Brown–Palmer family. This meant that one or two members would be occulted from time to time, just as the Moon occults ordinary stars. By noting the moment of occultation of a point-like radio-source and the moment of its reappearance, the time measurements together with the known motion of the Moon would determine the absolute position of the source in the sky to within an accuracy of only a second or two or arc, enormously better than any other method available in 1961–62, and amply good enough for an optical identification to be made. The particular member of the family of chaps on which Hazard set his eye was the radiosource 3C 273.

After testing on the 250 foot radiotelescope at Jodrell Bank that the method would work, Hazard left England to take up a research post at the University of Sydney, where there was no suitable radiotelescope for continuing the project. The suitable telescope was the 210 foot instrument at Parkes, N.S.W. But this telescope belonged to the Commonwealth Scientific and Industrial Research Organization (C.S.I.R.O.) and at that time a war in which no prisoners were taken was being fought between Harry Messel, Head of Physics in the University, and Taffy Bowen, Head of Radiophysics at C.S.I.R.O. Since Hazard had joined Messel's staff, there seemed no prospect that he would be able to use Bowen's telescope.

By a quirk of fate, however, John Bolton had just then changed his job, from being Head of Radioastronomy at Caltech to being Director of the C.S.I.R.O. radiotelescope at Parkes. Despite the war between Bowen and Messel, through the good offices of Joe Pawsey it was arranged for Hazard to have access to the telescope, with the work to proceed in collaboration with members of C.S.I.R.O. In taking this broad-minded step, Pawsey and Bolton opened the way to the great discovery then pending, and which would sweep over the whole scientific world a few short months later.

So it came about that Hazard, together with M.B. Mackey and A.J. Shimmins, obtained the position on the sky of 3C 273 to within an accuracy of one or two seconds of arc. It was then a simple matter to see from the Palomar Sky Survey that a 13th magnitude starlike object was at the position of 3C 273, and to obtain a low dispersion spectrum of a 13th magnitude object was by American standards an easy job. So what to do next?

I do not think Cyril Hazard's friends, or even Hazard himself, will quarrel too much with me when I say that Cyril does not rank at the summit of those who are given to precise office-administration. I have a never-to-be-forgotten vision from 1967, or thereabouts. The scene is the offices of the Arecibo radiotelescope in Puerto Rico. By this time Hazard has a job there, and Tommy Gold is the Director. Geoffrey Burbidge and I are on a visit, and Tommy is showing us over the telescope. After climbing for what seem hours over the huge structure, we go into the offices, and Tommy takes us along to see Hazard. Cyril is not there, however, but scores of rolls of pen-chart records are there alright. Not neatly rolled and stacked. Pulled open, scattered over the whole office like toilet rolls over a football field at the end of a big game. Except that the ceiling was not papered with them, they were everywhere. Tommy gazed around the office aghast, and then said in a small, hushed voice:

> 'To think I have paid tens of thousands of dollars for this.'

Back now to early 1963. For the reason which my anecdote makes clear, John Bolton perceived that if anything immediately decisive was to be done with the Hazard–Shimmins position of 3C 273 he had better do it himself. At that time, Bolton had not yet developed the proficiency in optical spectroscopy he was to acquire a few years later, otherwise he would no doubt have gone to the nearest suitable optical telescope and done the rest of the job himself. Instead it seemed natural to send the position to an astronomer in one of the American observatories. The choice was a fateful one, because, as it was to turn out, 3C 273 was the ideal object on which to make the crucial breakthrough.

It is a matter of history that Bolton sent the 3C 273 position to Maarten Schmidt at the California Institute of Technology. I think it was in January 1963 that Maarten Schmidt and Jesse Greenstein gave a joint paper in New York in which they sought to revive the pre-1951 concept of radiosources as local nearby stars. It must therefore have seemed to Schmidt that the 13th magnitude starlike object at the Hazard–Mackay–Shimmins position was a fine example to prove that the pre-1951 concept was correct, at any rate for a selected fraction of radiosources.

If one excepts the theoretical approach of Willy Fowler and myself, the situation was that in January 1963 nobody had any inkling of what was about to befall within the space of only a few weeks.

When Maarten Schmidt obtained the spectrum of 3C 273, the situation was drastically changed. Among the strongest lines were three which I will denote by β, γ and δ. While the wavelengths associated with these lines, λ_β, λ_γ, and λ_δ were not the laboratory wavelengths of the lines Hβ, Hγ, Hδ of the Balmer series of hydrogen, the ratios λ_β:λ_γ:λ_δ were exactly those of the Balmer series. This could mean only that the spectrum was redshifted by an amount that was determined to be about 16 per cent.

When Maarten Schmidt communicated this result to Jesse Greenstein it was I think only the work of a few days to interpret the spectrum of 3C 48. Within a month or so Schmidt and Greenstein had written a paper for the *Astrophysical Journal* which has since become a classic. It was in this paper that arguments against the redshifts being of gravitational origin were first given. As I indicated before, nobody has succeeded since in disturbing those arguments.

Nor did anybody in 1963–64, except J. Terrell, question the cosmological origin of the redshifts. Since I was especially happy myself to find that the objects whose existence Fowler and I had predicted really did exist, and apparently on an even larger scale than we had dared to hope, I was in no mood to think that the redshifts might have an origin otherwise than in the expansion of the universe.

It was soon realized that quasars are variable in their output of visible light. This could be seen by comparing recent observations with old patrol plates of the sky. The time-scale of the variability — several years or decades — was no embarrassment. It was otherwise, however, when in 1965 W.A. Dent discovered changes in a time-scale of about a year for the radio emission from both 3C 273 and 3C 279 (another quasar which by then had been discovered). Fluctuations of radio output were much harder to understand than optical fluctuations.

By 1966 the difficulty had become acute. Time variability seemed to require emission of the optical continuum light to be from regions not more than about a tenth of a light-year in diameter. If the redshifts were due to the expansion of the universe, the great distances then involved forced the light intensity within such regions to be so enormous that the light interacted destructively with the electrons that generated the radio emission. This problem, discussed in 1966 by Wallace Sargent, Geoffrey Burbidge and myself, has grown worse with the years, as ever more rapid examples of variability have been discovered.

Models which sought to evade the difficulty were put forward, notably by M.J. Rees and L. Woltjer. I have always believed myself that these models, while kinematically attractive, were physically incomplete. They assume that a system of particles can expand at relativistic speeds without the random energies of the particles being decreased adiabatically. For this to be so, a continuous internal source of random energy would be needed, and the models never specified the source in a way that seemed to me to have a satisfactory physical basis.

By 1966, there were further causes for disquiet. Two or three quasars were discovered close in their positions on the sky to nearby bright galaxies. If the quasars were really at great distances, as the cosmological interpretation of their redshifts required, these similarities of position would need to be fortuitous. One could easily see that the odds were against this being so.

In the years immediately following 1963, one or two quasars with very large redshifts were discovered, so large as to bring the edge of the so-called Lyman continuum of hydrogen, which occurs at 912 Ångströms in the laboratory, into the observable region of the spectrum. If the quasars in question were really at cosmological distances, and if there were any appreciable quantity of neutral hydrogen along the line of sight from the quasar to the Earth, the quasar spectrum in the Lyman continuum would be blotted out. Since the Lyman continuum was not blotted out, an opportunity to establish the cosmological nature of the redshifts had been lost. Either there was scarcely a vestige of neutral hydrogen in intergalactic space, or the redshifts of the quasars were not due to the expansion of the universe.

It is strange to contemplate that Geoffrey Burbidge and I thought the evidence sufficient in 1966 to question the cosmological theory of quasar redshifts. But this was before the present sociological trend toward extreme conformity produced a situation in which few dare question the cosmological theory, even though the evidence against it today has become overwhelmingly strong.

We managed to secure publication for the paper we wrote, *mirabile dictu* in the *Astrophysical Journal*. To do this nowadays you would need to have Cromwell's army standing behind you. The explanation lay in our friendship with the then Editor, S. Chandrasekhar. Faced with a choice between our friendship and the reputation of his journal, Chandra, after much thought, and to his great credit, gave the thumbsdown to the journal. So our paper stands there in black-and-white, an eternal disgrace to the Chicago University Press. In it we argued for the simple Doppler shift, theory (2) set out at the beginning, likely enough the correct theory as we shall presently see.

The Controversy Develops

Although the *Astrophysical Journal* was thus forced, almost at pistol point as one might say, to publish a correct theory, the journal swallowed its medicine very slowly. During the year which it took to publish our paper, Geoff Burbidge and I were often wakened in the dead of night by the telephone. A mysterious voice threatened us with the ruin of our scientific reputations. In my own case the threat proved to be alarmingly true. Over the years from 1970 to 1972, when I was involved contemporaneously in many struggles, the quasar issue proved a wonderful gift to my opponents, an obvious demonstration that the old man had lost whatever modest powers of scientific judgement he might once have possessed.

As the years rolled by, however, every one of the points Burbidge and I had raised became stronger and stronger. Time variability in the radio emissions from quasars became more and more rapid, to a point where one would have thought sheer commonsense would have brought the controversy to its appointed end. Quasars of large redshift were discovered in considerable numbers, many without the Lyman continuum being chopped away by intergalactic absorption. More and more curious associations between quasars and galaxies were discovered, especially by Halton (Chip) Arp.

The conformists were unlucky in having Chip Arp against them. If my memory serves me correctly, Chip had been No. 3 in the ranking of U.S. fencing masters, so that, although greatly outnumbered, he was able to skewer his opponents three-or-four at a time. However, most of the strange things which Arp discovered were

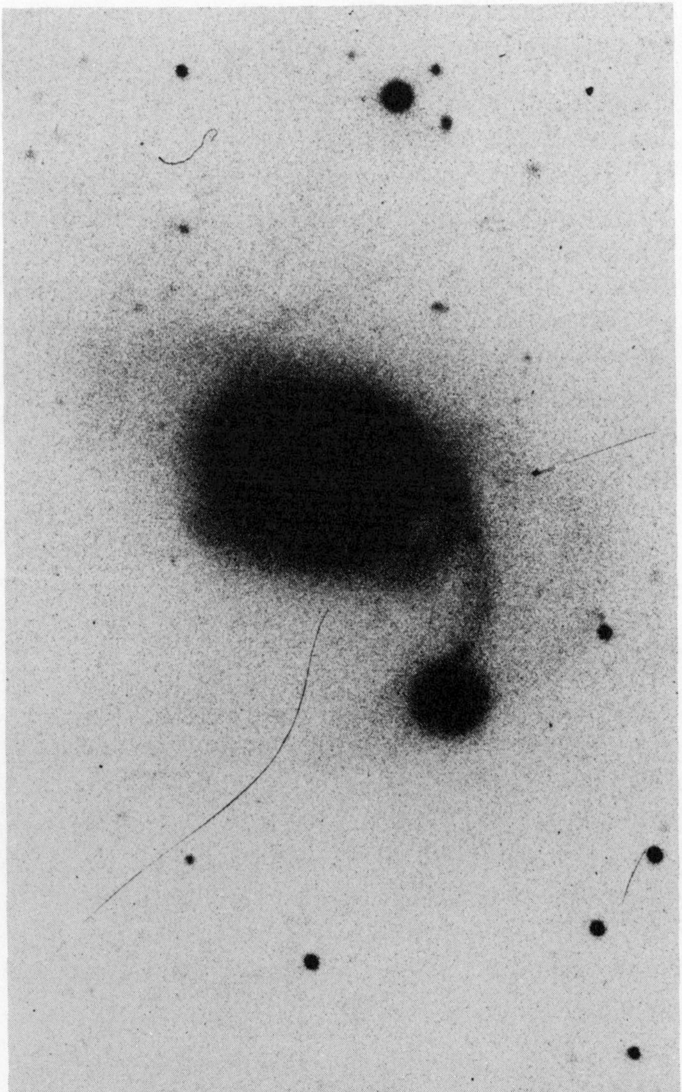

Figure 1
The galaxy NGC 7603 appears connected to a satellite galaxy by both an arm and an arc of luminous material to the right of the arm. Photographic negative.

hard to quantify, and it was this difficulty which permitted the controversy to continue.

Figure 1 is an example in miniature of Arp's discoveries. The large galaxy, NGC 7603, has a redshift of 2.6 per cent in its spectrum lines, whereas the satellite galaxy has a redshift of 5.3 per cent. According to the usual cosmological interpretation of redshifts the main galaxy would be at a distance from the Earth of about 300 million light years, whereas the satellite would need to be at a distance of about 600 million light years, which is to say the satellite would not be a satellite at all. It would have to be a much more distant independent galaxy that just happened to lie in the background. This would be a monstrous coincidence in view of the manifest fact that NGC 7603 is connected to its satellite, not just by one arm, but by an arm and an arc which intersect bang on the position of the satellite.

Of course there is a very large number of galaxies, and if one were to consider the full ensemble of galaxies some curious chance associations of position on the sky would undoubtedly be found. But there are only a few thousand galaxies as bright as NGC 7603, and there are only about a hundred that show the measure of internal disturbance to be seen in Figure 1. So the peculiarity of NGC 7603 cannot be gainsaid by an appeal to the enormous total number of galaxies. The peculiarity remains.

Figure 2 illustrates the difficulty of quantifying cases where quasars are found lying near bright galaxies. One can conceive of accidental associations, but if the quasar in Figure 2 (discovered by A.N. Stockton) is accidentally positioned, what a very strange clump of galaxies is involved in the accident! Because of the

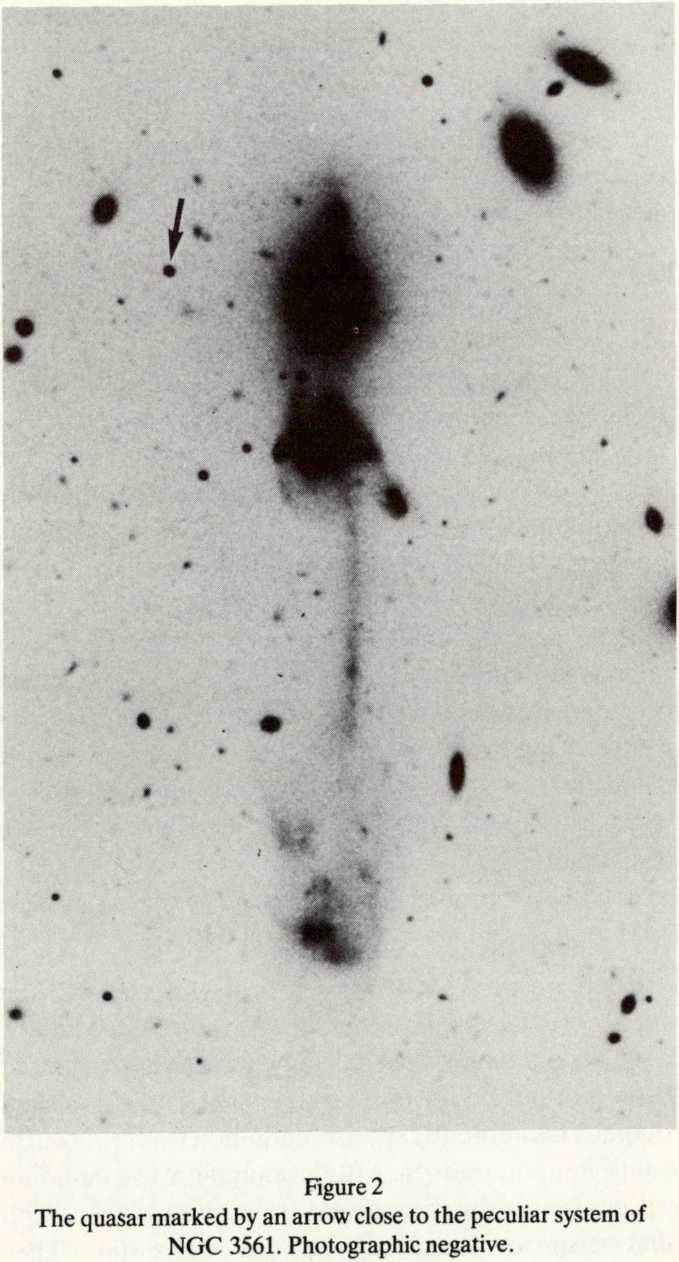

Figure 2
The quasar marked by an arrow close to the peculiar system of NGC 3561. Photographic negative.

difficulty in setting numbers to situations like Figure 2, I have preferred myself a different form of evidence, a form for which clear-cut calculations can be made.

Before I come to this other line of evidence, which has eventually turned out to be decisive, let me pause for a moment to answer a question that is sure to puzzle scientific historians of the future. What is it that has given conformists the confidence to ignore so many facts and arguments?

What really is a quasar? Is it a compact object with a certain kind of spectrum? Is it a compact object with a very large redshift? The disposition of astronomers has been to distinguish quasars by their spectra and not by the amounts of their redshifts, a criterion which permits *some* quasars to have redshifts of cosmological origin. There are quasar-like nuclei at the centres of a number of galaxies, and these certainly display redshifts that have their origin in the expansion of the universe. No person of non-conformist views that I have ever met has doubted this would be so. Indeed Willy Fowler and I predicted the general phenomenon even before quasars were discovered. What the conformists did was to find a few examples of this expected type, and then they asserted without proof that *all* quasars were of the same kind.

What is a wasp? If we were to answer this question in analogy to the situation for quasars, we would say that a wasp was an insect with certain characteristic coloured bands around a body with a characteristic 'wasp-waisted' shape. This apprently specific definition would of course admit imitation wasps. After examining a few imitation wasps, the conformists would be led into the assertion that wasps are harmless insects without a sting. They would then proceed to tear down with their bare hands the nests of genuine wasps, with results I find gratifying to contemplate.

Back to the Allen–Palmer–Hanbury Brown list of strange chaps. The radiosources 3C 286 and 3C 287 were both strange chaps, both quasars, and being adjacent to each other in the 3C catalogue they were adjacent sources on the sky in the third Cambridge survey. The number of quasars in the 3C radio survey was such that purely by chance one would expect such juxtapositions to occur in a handful of cases. So the mere adjacency of 3C 286 and 3C 287 was not a matter of any surprise. The surprise, a mild one at first, came when I took a look at their radio and optical properties. They were similar both in radio flux and optical magnitude, and they were not too dissimilar in optical colour either. The hint was that these two quasars might have shared a common origin. Yet they were about 5° apart on the sky, 3C 286 had a redshift of 0.85 and 3C 287 a redshift of 1.05. If redshifts arose from the expansion of the universe, there could be no such difference for quasars with a common origin. So either the observed similarities in radio and optical properties were fortuitous or the cosmological theory was wrong.

In 1965, Allan Sandage had startled the astronomical world by the discovery of objects that were optically similar to quasars but from which no radio emission had been detected. His name for these objects, quasi-stellar galaxies, did not pass into general use, however. The objects took on the general name of quasars, with the concession that quasars need not be radiosources. Indeed, a considerable fraction of the fifteen hundred or so quasars presently known are not radiosources, and complete surveys of limited areas of the sky down to about the 20th magnitude have shown that the majority of quasars are 'radio quiet'.

Sandage's discovery, taken with the blue colour of the quasars then known, suggested that a search for quasars be made in the available catalogues of blue starlike objects. The Tonanzintla catalogue quickly yielded a quasar which became designated as Ton 156, with a redshift of 0.549 and a visual magnitude of 16.4. Then in 1972 it was noticed by A.N. Stockton that the neighbouring member in the catalogue, Ton 155, was also a quasar with a redshift of 1.703 and a visual magnitude of 16.9. So here was another case of a quasar pair with different redshifts, and unlike the wide 5° spacing of 3C 286 and 3C 287, Ton 155 and Ton 156 were a mere 35 arc seconds apart on the sky.

If this discovery had been made in 1965–66 there might well have been no quasar controversy, since in the fluid situation of those years this one case alone could very well have destroyed the cosmological theory at a single blow. Even today there have been only 15 or so quasars found in the Tonantzintla catalogue. Although this catalogue covered but a fraction of the sky, the chance that two out of fifteen of its members would lie by accident within 35 arc seconds of each other was very small — I estimate it to be about 1 in 1000. My calculation has been made conservatively. A more ruthless attack would have yielded a still smaller probability. Yet the conservative result is sufficient, since it is usual in science to dismiss controversial theories that require fortuitous coincidences with probabilities of less than about 1 in 300.

Although not with the tiny spacing of Ton 155 and Ton 156, Ton 153 and Ton 157 are quasars also, with redshifts of 1.02 and 0.96 respectively. Here then, is a physically-associated cluster of quasars, and since they have different redshifts it is manifest that the cosmological theory is wrong. In 1965–66 the situation would have appeared overwhelming, but by 1972 the conformists were baaing so loudly all over the hillside as to drive everybody out of their wits.

The trouble with conformity is that the process has strong positive feedback. The baaing starts up at a volume low enough to permit stronger-minded animals to think for themselves without too much trouble. Progressively, however, we break down one-by-one, losing all power of sensible judgement, to the point where we can do nothing but add our own baaing to the uproar, which eventually rises to such monumental proportions that nothing remains for the flock except the butcher's shop.

Table 1 gives a number of quasar pairs that have been discovered over the years. Except for 0957+561, which may well be a single quasar made to look double by a gravitational lens effect, these nearly identical positions on the sky must be due to chance if the cosmological theory is to survive. The probability that chance would throw up the associations is given in the last column of the table. Except for OL 564, the most recently discovered pair,* the probability estimates are mine, calculated I believe in a fair way.

* F.N. Owen, B.J. Wills and D. Wills, *Astrophysical Journal*, 235 (1980), L57.

Table 1

Coordinates (R.A., Dec.)	Name (Catalogue)	Visual Magnitude	Redshift	Separation (arc sec)	Probability of chance association
0254−334	(A) PKS	17.0	1.915	54	2×10^{-2}
	(B)	16.0	1.857		
0957+561	(A)	17.0	1.405	5.7	Gravitational lens
	(B)	17.0	1.405		
1038+528	OL 564	17.5	~ 0.68	33.5	$\sim 10^{-2}$
		18.5	2.30		
1318+290	TON 155	16.9	1.703	35	10^{-3}
	TON 156	16.4	0.549		
1721+343	4C 34.47	16.50	0.206	60	2×10^{-1}
	No Catalogue Name	19.0	1.80		
2143−156	(A) PKS	17.50	~ 0.70	78	2×10^{-1}
	(B)	18.50	2.055		

I have remarked already on the scientific convention whereby controversial theories requiring fortuitous coincidences with probabilities less 3×10^{-3} are discarded. Of the pairs listed here only the Tonanzintla quasars yield such a result. Each of the other pairs has a probability larger than 3×10^{-3}. Hence if we agree to discuss each pair separately, and if we agree on a strict taboo where in the discussion of a particular pair the others are not mentioned, and are considered not to exist, and if we forget about Ton 155, 156, the cosmological theory survives the standard scientific convention. Baa! A better procedure of course is to multiply the probabilities given in the final column of the table, obtaining 10^{-8} as the measure of the coincidences necessary to bolster the cosmological theory.

Let us go on now to discuss the decisive evidence against the cosmological theory, beginning as before in an anecdotal vein.

Cyril Hazard Again

When in 1966–67 I was setting up the Institute of Astronomy at Cambridge, it was represented to me by those who were generously providing funds that it would be appreciated if I could do something towards reversing the so-called brain drain to the United States. Although in this endeavour I was only modestly successful, I did manage to persuade Cyril Hazard to return to the U.K.

Cyril claimed to begin with that he had lots of data 'to work up', which in view of the state of the office I had seen in Puerto Rico I could well believe. It needed no great gift of prophecy, however, to foresee that before long Cyril would be wanting a radiotelescope wherewith to accumulate more rolls of pen-chart records. But we had in Cambridge (or were then acquiring) about 2 million pounds worth of radiotelescope, built for the most part at the taxpayer's expense. Although these telescopes were under Ryle's administration, Hazard has such an easy-going personality, without a trace of my own astringency, that there seemed no reason why he should not have collaborated peacefully and fruitfully with Ryle's group of radioastronomers. It was not to be, however. I cannot say why Hazard's approach to the Ryle group failed, because I thought it tactful not to become involved myself. Nor did Cyril himself ever say too much about it. The objective fact was that before long he became discouraged.

Since Hazard was a product of Jodrell Bank, it seemed likely that an approach to Lovell would be more successful, and at first it appeared as if it might be. Lovell agreed to give Hazard access to the big 250′, subject to the proviso that his programme should not overlap work going on already at Jodrell. So pretty soon Cyril and I were driving the 150 miles from Cambridge to mid-Cheshire in order to explain to Lovell what Cyril wanted to do. Pretty soon too, we were driving the 150 miles back again in the knowledge that Jodrell was already at work on the same idea. A month later we repeated the identical performance. Likely enough we would have given the Jodrell connection a third shot if Cyril had not then come up with an interesting device.

To understand the subtlety of Hazard's device it is necessary to return to the blood-letting that was going on in Australia. Quite apart from the battle of Titans between Taffy Bowen and Harry Messel, there had been a profound disagreement within Bowen's own group at C.S.I.R.O. It was a disagreement by no means special to Bowen's group, or to Australia. The same disagreement existed everywhere. There was a general, and valid, difference of opinion between those who preferred to build dish-like tracking telescopes in the style of the 250′ at Jodrell and those who preferred to establish much larger but comparatively immobile collecting areas on the ground in the style of the radiotelescopes at Cambridge. Indeed, the difference between Cambridge and Jodrell reflected precisely this dichotomy, a difference that was manageable when it was separated by 150 miles, but which could not be contained within the same department, unless funds were available to support both.

In the early days of Australian radioastronomy, C.S.I.R.O. did indeed have the funds to support both, but as radiotelescopes spiralled in their costs a stage was reached when, for a period of several years at least, one side or other had to lose out. With the decision to build the 210′ Parkes telescope, it was Bernard Mills, the builder of renowned Mills Cross, who lost and it was John Bolton, a firm believer in precision steerable instruments, who won.

Realizing that Mills was not pleased by the situation, Harry Messel saw a great opportunity to explode a personnel bomb in the very centre of Bowen's own department. Messel succeeded in attracting Mills from C.S.I.R.O. to the University of Sydney, on the promise that he would provide money for a new bigger Mills Cross.

Harry Messel was a phenomenon previously unknown to science. He had gone from Canada to Australia, accepting the post of Head of Physics at Sydney University, a post which other physicists had taken a brief look at and had then instantly shied away from. The trouble was that the Physics Department was run down, the University (with the best will in the world) had no money to rebuild it, and owing to the peculiar way in which Australians finance their State Universities there was no prospect of obtaining federal funds either. An ordinary mortal plunged into this dispiriting situation would have sought to acquire goodwill by cringing and wheedling. Not so Harry Messel. Within a few weeks of arriving in Australia Harry had succeeded in offending every establishment figure throughout the length and breadth of the continent. Nor was the offence mild. It was war to the death on all sides. The fracas with Taffy Bowen was but a drop in the ocean.*

* It seems to have exacerbated the Messel–Bowen relationship in a curious way, that whereas Harry was a superb water-skier Taffy was a superb yachtsman, so that their paths kept crossing even in the middle of Sydney Harbour.

Yet, astonishingly, Harry Messel had succeeded. When I visited his Department for a couple of weeks in November 1962, physics was booming. Harry's own office was palatially laid out with a large expensive desk and with an enormous bearskin adorning the wall. With his feet high on the desk, a cigar in his mouth, Harry roared at me

> 'Christ man', in a voice like a million corncrakes, 'when I first came here I had to wear fisherman's waders to ward off the snakes'.

Harry had done it by foraging outside academic circles. Australian businessmen understood him and liked him. Harry's money was raised the hard way, commercially. His method of funding the department's scientific projects was commercial also in its style. Like a producer who begins shooting a film as soon as he has raised what is called 'front money', so Harry started to build a scientific instrument as soon as he had the front money for it.

It was a fine system, so long as you had an iron-strong constitution, which Harry had. Even so, Harry started so many projects that the ever-mounting problems of securing 'end money' had caught up on him by the time of my visit to Jodrell with Cyril Hazard. What Cyril now told me was that Harry had just about bust himself in securing the end money for the new Mills Cross, and that nothing remained in the kitty to buy the computing facilities that would be needed to take proper advantage of the data that was beginning to roll off the Cross.

The days were passing when radioastronomers would simply publish the positions of sources on the sky, like those in the left-hand column of Table 1, leaving it to the optical astronomers to skim the cream from their work. The days were also passing when radioastronomers would laboriously measure positions on prints of the Palomar Sky Survey. A far better method had been invented. The positions on the sky of a very large number of stars, well-known in astronomical catalogues, had been punched onto computer cards. After reading the cards into a computer, it was a trivial matter to transfer the star positions onto magnetic tape. Many copies of the first tape were then made and distributed around the world to interested parties. If you acquired one of these tapes you could feed the star positions into your own computer, provided your computer was big enough to accommodate all the data. This was the starting point of what Hazard was proposing to me.

Suppose next that you have the position of a new radiosource and that it is your wish to find out if an optically-visible object exists at precisely the same position. Your reference material is the Palomar Sky Survey, which consists of prints (or glass plates) showing optically-visible objects down to a very faint limiting magnitude, each print covering a 6° × 6° area of the sky. You supply the computer with the position of the new source and you instruct it by a program to do the following:

(1) Find all the stars (on the magnetic tape) which lie within, say, 20 minutes of arc of the specified radiosource position.
(2) Print out all these stars in the form of a map centred precisely at the radiosource position, and do so on exactly the same scale as that of the prints of Palomar Survey.

To effect this program you require suitable printing equipment, as well as the computer itself. The printing equipment must have an unusual property. Instead of the map of star positions being represented by marks on opaque paper you arrange for them to be marks on a translucent material. You now overlay the translucent computer map onto the appropriate Sky Survey photograph, in such a way that the marks printed by the computer fall exactly on the actual stars to which they correspond. All you have now to do is to take a detailed look at the very centre of your map. If there is any visible object on the Survey photograph at that central position your aim is achieved, that is the radiosource.

Although easy to conceive of, this plan involved fairly sophisticated equipment by the standards of the late nineteen-sixties. A sum of money of the order of a quarter of a million pounds was involved. According to Cyril Hazard, the situation in Sydney was that Harry Messel had so many demands on his cash flow that he could not whistle up such a sum at a moment's notice. Since it seemed to me that the only problem for the Institute computer would be a modest amount of new specialist printing equipment, I agreed without question to Hazard's proposal for a collaborative project with Sydney radioastronomers working on the new Mills Cross. To wit, they would supply the radio positions and we would deal with the computer problem in order to obtain the required optical identifications.

Pretty soon, however, the computer experts were telling me that we needed a good deal more than a bit of specialist printing equipment. We needed a further block of high-speed store, costing about £50,000. I had only a vague second-hand knowledge of the state of

Harry Messel's cash flow, but I knew for an absolute certainty that mine would not stand up to such a demand.

At that time, 1967–68, a sum of £5 million for Ryle and Lovell was being talked about at the Science Research Council, and in comparison with £5 million a sum of £50,000 for Hazard's project looked very small. I therefore contacted Jim Hosie who was in charge of the Astronomy and Space Division of the Science Research Council. Hosie had taken a double-first in the Mathematical Tripos at Cambridge, and so he had no difficulty with numbers, as I shall relate in more detail at a later stage. Being also a Scotsman and a Highlander to boot, he could see that securing a half-share for £50,000 in a project where one's partner had spent perhaps £1 million was in the nature of a bargain.

There was then a six-months delay in the delivery of blocks of core storage, which was also about the time it would take to prepare a grant application to the Science Research Council, and for the application to receive consideration by the three tiers of the Council's committee structure. Together, this amounted to a year's delay, by which time the opportunity might well be gone. So I decided to cut a sharp corner. On the strength of Hosie's nod, and (if the truth be told) in cognisance of the fact that I was the Chairman of one of those three Science Research Council committees, and a member of the other two, I ordered the block of core immediately.

It was a serious administrative mistake. I overlooked a decision which had been taken a year or so before I joined the Science Research Council. Grants for radioastronomy in Britain would be confined to the Ryle and Lovell groups. Of course, the decision had been taken to avoid radioastronomy outfits mushrooming throughout the U.K. with demands for sums in the millions, not to stop a comparatively small-scale affair like mine. Even so, when the grant application eventually came before the Research Council committees, the letter of the previous decision was quoted against me. The issue was seen as one of deep principle by both the Jodrell and the Cambridge radioastronomers. If I were allowed to breach the rule, it would be hard to stop others from breaching it. Nowadays, having lived a decade outside the halls of academe, I would answer this argument with Harry Messel's nerve-shattering laugh, but in those days I was far too mild in such things, and in the event I allowed myself to be put down.

By this time, the IBM Corporation had wheeled in the block of extra core, had buttoned it up to our computer, and had reported to me that everything was in excellent shape. Ironically, at my first business contact with Harry Messel I now found myself in exactly the kind of situation in which Harry himself would have revelled, but in which I did not.

There was just one trace of a silver lining to the cloud. Besides facilitating Cyril Hazard's project, the extra core would increase the computer throughput by about fifty per cent. So we would have spare computer time to sell, if we could find a purchaser. On the strength of this pious hope we borrowed the money. By the time I left Cambridge in 1972, more than £35,000 had been repaid. Most of the repayment money came from the Medical Research Council, which thus had a hand in the remarkable discoveries I am about to relate. There was also a stage where we managed, peculiarly enough, to sell time back to the mighty IBM Corporation itself.

The day eventually came when Cyril Hazard was standing there in my office holding out a stack of polaroid pictures.

> 'You're not going to paper my office with those things', I began.

Cyril grinned in his usual way and held out a picture which he had shuffled from the pack, saying,

> 'Take a look at this one.'

To the eye it looked liked a double-star with a separation of about 5 arc seconds. Cyril pointed to the brighter of the two components, adding, 'That's the source'. The object was blue, making it a likely quasar. The other member of the apparent double-system was blue also. Were they both quasars, and if so what were the redshifts?

The fainter member was at the 19th magnitude. To obtain its spectrum would not be a simple job by the standards of the day. No equipment then available to British astronomers would suffice. So, through the good offices of Margaret Burbidge, contact was made with Joe Wampler at the Lick Observatory. When the results eventually came through, they were just as I had hoped. Both components were indeed quasars, the brighter with a redshift of 0.436, the fainter with a redshift of 1.901. Although the results were not published until 1973,* I

* E.J. Wampler, J.A. Baldwin, W.L. Burke, L.B. Robinson and C. Hazard, *Nature*, 246 (1973), 203.

was aware of how the investigation had gone by the time I came to leave Cambridge in the autumn of 1972.

To my still-innocent mind it seemed that the cosmological theory must now surely be stone dead. Since the redshifts were different, the cosmological theory required the juxtaposition to be fortuitous. The probability of such a fortuitous juxtaposition appearing among the sprinkling of new quasars which Hazard had found in the data from the Mills Cross (referred to nowadays as MC 2 quasars) was easily calculated to be about 1 in 1000. The details were as follows:

Coordinates	Name	Visual Magnitude	Redshift	Separation (arc sec)	Probability of chance association
1548+114	(A) 4C 11.50	17.0	0.436	5	10^{-3}
	(B) MC 2	19.0	1.901		

Readers in possession of their faculties will be astonished to learn that this result did not settle the controversy. The astronomical world continued exactly as if nothing had happened. By this time I myself had had enough of it, so I fear that over the next few years I left the running to the resolute Geoff Burbidge and to the doughty Chip Arp. Only once have I emerged upon the stage, when the American Astronomical Society did me the honour of inviting me to give its annual lecture in memory of Henry Norris Russell. I spoke on quasars in rather the astringent tone of this essay, and gratifyingly I think the audience half-believed me for the space of an hour or two.

Shortly thereafter, I had a letter from the Secretary of the Society to say that the A.A.S. had made an arrangement with the *Astrophysical Journal*, according to which the Journal would publish each year's Russell Lecture, and would I therefore please submit a manuscript to the Editor. I did so, sending an exact verbatim copy of what I had actually said. Pretty soon I had a letter from the Editor revealing a heart-rending situation. It seemed that the Chicago University Press, having read my lecture, was in a state of deep shock. Would I, the Editor wanted to know, in order to save the Press from further acute pain, be willing to emasculate the lecture? I wrote immediately to the Secretary of the A.A.S. suggesting the Society should sort the matter out with the *Astrophysical Journal*. The Editor continued to send letters for a while, repeatedly begging for emasculation, but I never did hear again from the A.A.S.

Cyril Hazard was aware that the halcyon days in which he was receiving precious information from the Mills Cross could not last. It therefore behooved him to wring as much as possible from what he had available. It was this need, this necessity, that finally broke the problem, up to the enormous odds which exist today against the cosmological theory. Instead of perpetually rushing from one source to the next, instead of attempting to process as many identifications as possible, Cyril came to savour each one of them like a fine wine. Which is to say, he began to look seriously at the fields which surrounded each source, finding in some cases a veritable halo of blue starlike objects. If the blue objects were quasars, then there could be little doubt that quasars occurred in clusters, as the case of Ton 153, Ton 155, Ton 156 and Ton 157 has suggested already. To decide the matter, spectra were needed. Nothing could be done towards securing spectra with instruments then available to us in the U.K., so once again there had to be an appeal to U.S. astronomers.

Before I left Cambridge in 1972, I was wont to include one or two of these possible quasar clusters in the lectures I gave in the United States, hoping somebody there would feel impelled to obtain the needed data. There was a particular cluster at coordinates 1146+111 that will assume a crucial role at a later stage of my story. But neither Cyril nor I succeeded in getting anybody interested. The trouble was that with respect to the techniques available in 1972 obtaining the spectra of faint quasars was a laborious business. It was one thing to take trouble for an actual radiosource itself, but to take much more trouble, and to consume much precious telescope time, on objects which might be there in the field surrounding a radiosource only by accident did not appeal to anybody.

The route towards cracking this problem had to follow a quite different course. Nobody could have conceived in 1972 that the critical optical telescope would not be in the U.S. It was to be a modest telescope then under construction at a site in northern N.S.W. Since I myself had a fair amount to do with that telescope in its early days, and since there were anecdotes and considerable in-fighting at its inception, the details of which will certainly not be found in the official records, it may have a little historic value if I interrupt the story of the quasar controversy to recount the story of the U.K. Schmidt Telescope.

If I allow my memory to wander back to undergraduate days, to lectures for the Cambridge Mathematical Tripos which I attended in 1935–36, I see a lecture room in the second court of St. John's College. Like so many nostalgic things, the lecture room doesn't exist today. It was over in the corner beside the library, and you reached it by the staircase on which Paul Dirac had his rooms. Also I see deep in memory's eye an undergraduate towards whom I used to nod, an undergraduate who had come up to Cambridge from Glasgow University. Between the summer of 1936 until 1965 I neither saw nor heard anything further of Jim Hosie.

After leaving Cambridge, Hosie had entered the Indian Civil Service, soon to find himself catapulted into a position of high responsibility by the outbreak of the 1939–45 war. At the termination in 1949–50 of the British administration in India he had returned to the U.K. to begin a new career. His abilities were such that by 1965 he had become the Director of the Astronomy and Space Division of the newly-formed Science Research Council.

Although I was to work quite a bit with Hosie in the period 1967 to 1972, he said remarkably little about his first career, as if he had walled-up memories of it in an inaccessible part of his brain. Only if one suggested eating a meal in a Chinese restaurant would he burst forth with the exclamation: 'Good God, no!' followed perhaps by the remark: 'You should see the things I have seen!' Like Great Aunt Ada Doom in *Cold Comfort Farm*, we never found out exactly what it was that Jim Hosie had seen in the woodshed.

The Institute buildings at Cambridge were completed in early August 1967. Within a month of that date, Jim Hosie walked into my office with the news that Hermann Bondi had just resigned the Chairmanship of the Science Research Council's committee on astronomy, which went in those days by the initials A.P.G.C. The purpose of Hosie's visit was to discover if I would assume the Chairmanship, a job which I had already done from 1959 to 1964 in the days of the Department of Scientific and Industrial Research (D.S.I.R.), the body which had preceded the Scientific Research Council (S.R.C.). Would I also become a member of the Research Council itself, and of the Astronomy Space and Radio Board chaired by Bernard Lovell? In the busy days of settling up the new Institute at Cambridge, these invitations were not entirely welcome, and if I hadn't been cursed by memories of the past I would probably have refused. Instead, I somewhat weakly said yes.

Money was then being broadcast by the S.R.C. all over the S. Hemisphere, as if Captain Cook himself had organized a paper chase of the stuff. None of this open-handedness was of the S.R.C.'s own making, however. There was money for the Cape Observatory which had been taken over by the S.R.C. from the Admiralty along with the Royal Greenwich Observatory. There was money for the Radcliffe Observatory in Pretoria, which Dick Woolley the Astronomer Royal had obtained from the old D.S.I.R. And there was a much bigger block of money (a 'pot of gold' as Hosie used to call it) now being allocated for future spending in Australia. The problem for the small Southern Hemisphere Review Committee consisting of Jim Hosie, Bob Wilson, then of Culham and now at University College London, and myself as

Chairman, was not just to make economies in this somewhat chaotic situation. We had also to consider whether the way we were going in Australia might lead later to British astronomers being seriously disadvantaged in their eventual use of the Anglo-Australian Telescope (AAT) compared to Australian users who would enjoy the geographical advantage of the AAT being sited in their own country.

The easiest way to reduce expenditure would have been to stop all activities in S. Africa, a course that would doubtless have commended itself to some people, but the Southern Hemisphere Review Committee did not travel to S. Africa on 27th April 1968 with instructions to examine race relations there. Our job was to report on what was best for British astronomy.

The first thing to be said about the S. African situation was that the Cape Observatory had long traditions. Many able British astronomers had indeed avowed that the achievements over the years of the Cape Observatory exceeded those of the Greenwich Observatory. A decision to close down the Cape Observatory could not therefore be taken lightly. Nor could we make the slightest scientific case for the S.R.C. to withdraw its support from the Radcliffe Observatory, for the overwhelming reason that the Radcliffe Observatory was at that time the most cost-effective part of British observational astronomy.

The Southern Hemisphere Review Committee recommended that the hitherto separated activities at the Cape and at the Radcliffe be joined, and that the worthwhile telescopes at both Observatories be moved to a first class-site in the north of Cape Province. We found during our visit that the South African Government was interested in joining this scheme and would pay all site-development costs, and this too we recommended. As I have explained, our job was to save British taxpayers money, not to make moral judgements. Our Report was eventually studied by those who could more properly make moral judgements. After duly considering such other questions it was decided to adopt our recommendations. It was also decided not to publish the Report lest it become a political football, an interesting shaft of light on the obscure corners of the bureaucratic mind.

The more important long-term effect of the Southern Hemisphere Review came from the other half of our remit, which was to give British astronomers a reasonable chance of a fair share of the large 150″ telescope (AAT) that we had begun constructing jointly with the Australians. The problem for us was that the instrument was to be in faraway New South Wales. I happened to know that an astronomer prominent then in the affairs of the AAT was going around the U.S. saying: 'Look now at these poor old Brits. Funding half a telescope at a distance of 12,000 miles. How crazy can you get?' I convinced Jim Hosie of this reading of the astronomical bush-telegraph, to the point where he was willing to aid and abet a devilish scheme.

I must explain that a very large telescope like the AAT only looks at a tiny bit of the sky, so tiny that you cannot possibly make a systematic survey of the whole sky with it. So unless you know already the position of the object you wish to observe before you point the telescope, the instrument is well-nigh useless. If you start hunting at random, the chance of finding something interesting is like picking out a needle in a haystack. This problem afflicted the old Mt. Wilson observers when they first built the 100″ telescope there. When the 200″ was constructed at Palomar Mt., a companion telescope of smaller aperture and of a remarkable design was built, the 48″ Schmidt telescope, named after Bernard Schmidt who first employed its optical system in astronomy. The design of a Schmidt telescope enables it to cover the sky in big chunks. The 48″ Schmidt at Palomar didn't penetrate out into space as far as the 200″, but experience showed that it could detect (in less detail) almost all the interesting objects that the Palomar astronomers wished to observe. What they did was to go systematically over the sky with the 48″ Schmidt, building a library of about 800 plates,* which showed the objects visible on the whole of the portion of the sky that was visible from Palomar. Thus the 200″ never had to operate blind.

The region of the sky around the southern celestial pole was not visible at all from Palomar, and so was not represented in the Palomar Sky Survey. On this region the AAT would be blind for all faint objects unless something was done about it. Indeed about a half of the sky visible with the AAT was either covered only poorly by the Palomar survey or not at all. It is difficult to understand how those who first planned the AAT were obtuse enough to have overlooked this quite crucial issue. No reworking of the agreement between the British and Australian Governments was possible, because the agreement was careful to specify that there be only *one* telescope. Somebody I think had played

* This number was duplicated, with one set taken in red light and the other in the blue.

tricks with that agreement, without quite understanding what they were doing.

The devilish scheme of course was that we poor old Brits should build a 48″ Schmidt based on the Palomar instrument, so giving us an advantage over the Australians.

Jim Hosie got to work with one of his financial juggling acts. He had a rolling five-year budget, with various sums assigned to the APGC in each of the five years. It was a crucial part of the game that any money which remained unspent in a particular year could not be passed on to succeeding years. If it was not spent it was 'lost', i.e. it returned to the Treasury. Not much was ever lost by Jim Hosie, however, even though the game was far from easy. The radiotelescope being built by Ryle, for example, was scheduled in Jim's rolling circus to consume about £1.5 m, with the £1.5 m distributed over about three years. If the bills which had been planned to come through in the first year didn't come through until the second year (a circumstance that could be critical to only a few days around the fiscal end of the year, 4 April) all hell was let loose. The money from the first year would then be 'lost', and would have to be 'found' in the second year. 'Finding' money in this way implied taking it away from other projects, and the baying of the sponsors of those other projects would inevitably be heard along the corridors of the S.R.C. for weeks on end.

A big factor in Jim's fiscal problems was the AAT itself. Quite against the spirit of the agreement between the British and Australian Governments, the AAT formed a part of the APGC budget. The treaty between the governments implied that the AAT would be handled at a level not much below the responsible Minister, in our case the Minister of Education and Science. It was already a step down from the Minister to the S.R.C., and it was another long step from the actual Council itself to the APGC. It was a bit like funding a major civil project through an old-style parish council. One advantage for the APGC in having the AAT expenditures appear in our budget was that it made our total outgoings look quite large, and since the annual growth rate we enjoyed at that time was calculated on our total, we were in effect earning compound interest on the AAT project itself.

Another aspect of Hosie's financial wizardry was that he had an uncanny knack of anticipating underspends, and of then slipping in reserve projects to avoid 'losing' money to the Treasury with whose officials he nevertheless maintained excellent relations. Hosie had been in the same year at St. John's College as Raymond Bell, the very same Raymond Bell with whom I had roughed it in 1936 in the Cuillin hills of Skye, the same Sir Raymond Bell who was now an influence at the Treasury. It is a mistake to think of the British old-school-tie-network as being entirely confined to Eton, Harrow and Winchester.

I asked Bruce Rule at Caltech, who had been responsible for the mechanical design of the 200″ on Palomar, to give us an estimate for a new 48″ Schmidt. After an hour or so of calculation with a slide rule he came up with £500,000. By mutual agreement, Hosie and I increased the amount by 50 per cent, which in our experience was about par for the escalation course where scientific instruments were concerned. Then Hosie went to work on his rolling budget. After a while he gave a quick shuffle to his pack of cards. Holding them out in a fan he asked me to pick one. I picked one and on it was written £750,000.

I have exaggerated only a little. It happened almost exactly like that, and the £750,000 was to prove just about the most accurate estimate of any instrumental project in my time at the S.R.C.

There were objections to the 48″ Schmidt proposal from both groups of British radioastronomers. Ryle didn't like it because it seemed too far away in the south, and Lovell didn't like it because it seemed to be taking away £750,000 from a large new radiotelescope to cost about £4 million which had already been agreed for Jodrell Bank. I well remember a set-to between Lovell and Hosie over it. Hosie produced detail after detail to prove that if the £0.75m didn't go to the Schmidt it would be lost to Ray Bell at the Treasury.

> 'But *you* can't use it, Bernard,' Hosie kept saying, 'it just can't be fitted into your timetable.'

Lovell had a particular laugh which I always envied him. He never used it except to express profound disagreement, and yet it was a pleasantly modulated laugh, never a nerve-splinterer like Harry Messel.

> 'Ah-ha!' laughed Lovell, 'but *you* can make figures come out to prove anything at all. We all know it. Even Hoyle there, whose scheme this obviously is, knows it.'

In truth, Jim Hosie with his training in the Mathematical Tripos, could make the figures in his rolling budget prove anything at all. But in this case

Hosie was right, and after a few more laughs Lovell conceded as much.

There was a lost of sporadic gunfire about the 48″ Schmidt, objections surprisingly from astronomers all over the world, which had an adverse effect on otherwise neutral members of the S.R.C. committees, to a point where it was becoming doubtful that Jim Hosie and I could carry the issue. Then, with the project balanced on a knife-edge, Dick Woolley joined the battle on our side. Woolley's intervention swayed the doubters and the day was won.

Bob Wilson, Jim Hosie and I had submitted our Southern Hemisphere Report to the S.R.C. shortly after our return from S. Africa in early May 1968. Yet it was not until the beginning of 1971 that we received the green light for the construction of the Schmidt. The time had gone not so much in arguing with Lovell as in the S.R.C.'s circulating committee system, and if anything in the least contentious were involved there would always be more than one such circuit. Everything was just about completed in the late spring of 1970, and Hosie had at last gone to the Treasury for final permission to spend the £750,000, when the Government changed. The need of the new Government to review its future financial commitments then took up the remaining months of 1970.

So two-and-a-half years had gone in breaking down the administrative impediments. What was to happen next demonstrated the astonishing speed with which the British people can move once the leaden boots of bureaucracy and tradition are removed. Both Hosie and I were determined the project was not going to be handled by a committee. We both wanted a key man to oversee and control the construction of the telescope. It was natural to begin by thinking of somebody at the Mt. Wilson and Palomar Observatories who had first-hand knowledge of the similar instrument on Palomar Mt. But of course you can't expect a good man to quit his job in order to build you a telescope, especially if it means a switch from the U.S. to the U.K. with an inevitable plunge in salary.

One day Hosie asked me: 'What do you think of Vince Reddish?' I knew Vince well, and partly because of this and partly because he was a theoretical astronomer, I was biased in his favour. But was a theoretical astronomer the right kind of person to control the building of a quite important and tolerably sophisticated telescope? I knew that even the Palomar astronomers had been in some difficulty to get their Schmidt to work properly in the old days. When I put these points to Hosie, he said: 'What about Galaxy?'.

'Galaxy' was an automatic machine for measuring astronomical plates. As well as determining the position of objects on a plate, which an American device called a 'Vasilevsky machine' would already do, 'Galaxy' was designed to measure the brightness (magnitude) of the object. The concept was due to P.H. Fellgett, who had left the Royal Observatory at Edinburgh in the middle of the construction to take-up a professorship at the University of Reading. A good deal of money had already been spent, so both the S.R.C. which had provided the money, and Hermann Brück, the Edinburgh Director, were anxious that 'Galaxy' should not abort. Probably only Vince Reddish on the Observatory staff had the mathematical knowledge to penetrate the subtlety of one of Fellgett's concepts, so it was Vince whom Brück had asked to complete the project. Now there is no more thankless job in the scientific universe than taking on somebody else's brainchild. Yet Vince Reddish had calmly picked up the pieces from the lab floor, and after what I was sure had been a tough and unremitting struggle had at last made 'Galaxy' work.

'O.K.,' I said to Hosie, 'it's Vince.'

Reddish began with what I thought to be a serious mistake. The designers of the Palomar Schmidt had anticipated flexure in the telescope's plateholder, and to compensate for this difficulty they had deliberately designed flexure into the tube of the telescope. It was a case of making one error cancel another. What Vince did was to deviate from the Palomar design* so as to stiffen the tube, so that he was then committed to avoiding flexure in the plateholder. If this could be made to work it would be an improvement, because a stiffened tube would give a greater versatility (as for instance in the use of an objective prism). But could it be made to work?

My disposition was to argue the matter, but then I decided the project was Vince's now, not mine, as I had always thought of it up to this point. It was right for me to mention my doubt, which I did, but the decision was up to Reddish.

Vince finished the telescope in the unprecedently short time of two years flat. I was privileged to see the finish, indeed to be the first person other than Vince

* Horace Babcock, the Director of the Hale Observatories, had kindly made the design available to us.

himself to see the finish. During the later stages of construction of the AAT, Jim Hosie and I arrived at the site on Siding Spring Mountain in mid-March 1973. Vince had already been there a month, struggling with the Schmidt. I knew it would be the plateholder. Hosie and I were at breakfast when Reddish came in one morning, his eyes red-rimmed with fatigue. After breakfast he said: 'Will you come along and take a look at the plate I got last night?' So instead of going off to bed, Reddish took me to the attractive little Schmidt dome alongside the AAT. Putting his plate on a viewing stand, he switched on the background light. Now he could contain his excitement no longer: 'Look!' he exclaimed, 'the images are in focus right to the plate edge'. And so they were. The plateholder bogy had been overcome.

The Schmidt dome had a special feature. It was fitted with blowers so powerful that when run at full blast you felt as if the place were about to take off. There had been a long-standing controversy among astronomers about exchanging air rapidly between the inside and outside of an astronomical dome. First put forward I believe by the French, the idea was to suck away heated air from the telescope, air that would otherwise rise up into the dome slit, so producing the effects of 'bad seeing', even if the atmosphere outside the dome happened to be good. Nobody until Vince Reddish had fitted blowers powerful enough to settle this controversy beyond cavil. The idea was right. You had only to stand at the telescope viewer in that roaring cockpit for two or three minutes to see the big difference the blowers made.

Vince Reddish has constructed other instruments since then, and always with speed and good judgement.

Reddish in my opinion is the finest builder of optical telescopes to have emerged in Britain since William Herschel. I can only hope his recent retirement from Edinburgh will not mean the end of a splendid run of successes.

In which the persistent Cyril Hazard and the doughty Chip Arp deliver the coup de grâce

I must now eat a fair-sized slice of humble pie, for it was Reddish's modification of the Palomar 48″ Schmidt that turned out to be spade which finally buried the cosmological theory of quasar redshifts, the stiffening of the tube to which I had objected in 1971.

An objective prism is a heavy thing, and since it has to be mounted at the upper end of the telescope tube, it puts a large shearing force on the material of the tube, causing it to flex unacceptably unless the tube is strongly built. The point of an objective prism is that its use yields a spectrum of every object in the field of the telescope, and it does so all at the same time, not one at a time as with an ordinary telescope.

Vince Reddish must have had his heart set from the beginning on the use of an objective prism, but his first job when the U.K. 48″ Schmidt came into operation was to organize the Sky Survey which had been planned in 1968. It was therefore some time before the objective prism could be fitted to the telescope.

Even the best-designed optical systems compensate in some degree for their advantages by disadvantages in other directions. While an objective prism on a Schmidt telescope has the advantage of yielding the spectra of all objects in the field in the same photographic exposure, thus gaining enormously in speed over an ordinary telescope, it has the disadvantage that the spectra for faint objects like quasars are of poor quality. Nevertheless, the quality is usually sufficient to distinguish a quasar from an ordinary star. The system is thus ideal for amassing information on the statistics of quasars, both with respect to the numbers of quasars at various magnitude levels, the so-called luminosity function, and to answering the question of whether there are significant large-scale variations of the quasar density on the sky.

When at last an objective prism was fitted to the U.K. Schmidt these were seen as the problems to tackle, and a number of valuable programmes have been carried through along these lines, by Malcolm Smith, by Vince Reddish, R.D. Cannon and M.R.S. Hawkins at Edinburgh, and by Ann Savage and John Bolton in Australia.

There was someone waiting in the wings of the stage with a very different idea, however. Who else but the persistent Cyril Hazard? There were still crucial questions to be answered about those old MC 2 radiosources, the ones dug out from Harry Messel and the Mills Cross. In particular, was that halo of blue objects around the radiosource at 1146+111 really a cluster of quasars? At long last, Hazard got his hand on a U.K. 48″ Schmidt plate that answered his question, a plate that would not have been possible but for the objective prism.

From a microscopic examination of the spectra a considerable fraction of the blue objects were seen to be indeed quasars. Equipped with this information it was then possible to persuade observers with access to the Anglo-Australian Telescope (by then completed) and to the large American telescopes to obtain high-quality spectra of the relevant objects. This was done by Chip

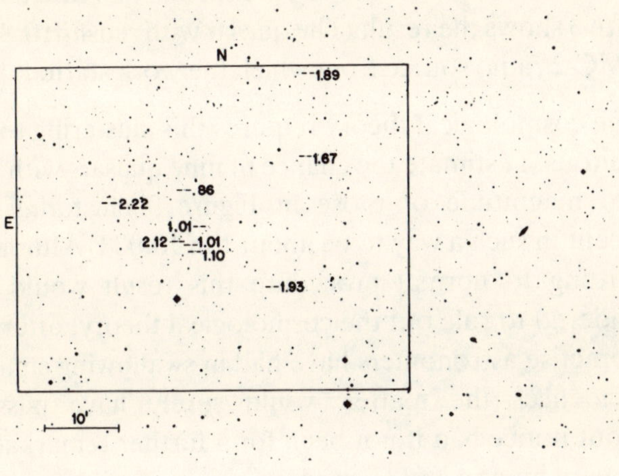

Figure 3
The cluster of quasars at position 1146+111. Two-aligned triplets can be found in this figure.
(From Arp and Hazard)

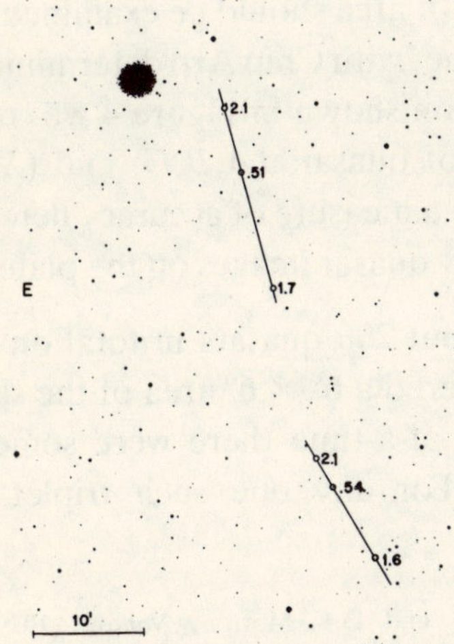

Figure 4
Two-aligned triplets of quasars near 1130+106.
(From Arp and Hazard)

Arp in the U.S. and by Don Morton in Australia.* Figure 3 shows the result. The quasar with redshift 0.86 is the MC 2 radio source from which the work started.

The cosmological theory requires this clustering to be fortuitous. I estimate the chance of nine quasars with the visual magnitude of those in Figure 3 clustering by accident in such a way to be about 1 in 10,000. Although according to normal procedure this result would be considered to rule out the cosmological theory, after all the practise astronomers have had in swallowing similar coincidences the matter would surely have passed without notice had it not been for a further remarkable discovery by Arp and Hazard.

Now that he had waited almost ten years for the precious plate of the 1146+111 region, Cyril Hazard was determined to treat it like a bottle of Mouton Rothschild. He was determined that every suitable square millimetre of area should be examined with care. When this was done, with Chip Arp determining redshift values, the situation shown in Figure 4 was uncovered. The two triplets of quasars (A,B,C) and (X,Y,Z) are collinear to within a measure of accuracy determined by the sizes of the tiny quasar images on the plate.

There were about 200 quasars in total on the whole plate,† which covered a $6° \times 6°$ area of the sky. Taking the quasars three-at-a-time there were some $(200)^3/3!$ distinct triplets. For any one such triplet choose a

* C. Hazard, H.C. Arp and D.C. Morton, *Nature*, (1979), 271; also H.C. Arp and C. Hazard, *Astrophysical Journal*, 240 (1980), 726.

† Down to magnitude 19.5, which is fainter than any of the quasars in Figures 3 and 4, considerably fainter in most cases.

particular member. The chance of *both* the other members lying within 10 arc minutes of it was $(\pi/6^4)^2$. Each triplet forms a triangle on the sky, and the number of such triangles having a largest side of the order of the line segments of Figure 4 was evidently

$$\sim \frac{(200)^3}{3!} \cdot \left(\frac{\pi}{6^4}\right)^2.$$

If we take the astronomical images to be defined to within 1 arc second then the triplets of Figure 4 are collinear to within an angle of about 0.1°. Since the average angle in a triangle is 60°, the chance of a triangle drawn at random having an angle less than 0.1° is 1/600. Thus the chance of obtaining one triplet at random like (A,B,C) or (X,Y,Z) was

$$\frac{1}{600} \cdot \frac{(200)^3}{3!} \left(\frac{\pi}{6^4}\right)^2 = 1.31 \times 10^{-2}.$$

In order of magnitude, the chance of obtaining two such triplets was therefore 10^{-4}.

The triplets of Figure 4 are not in the immediate neighbourhood of 1146+111. When Arp and Hazard showed Figure 4 to Bill Saslaw, Saslaw returned to the cluster of quasars near 1146+111 and proceeded to find two more alignments like (A,B,C) and (X,Y,Z). These further triplets are on Figure 3, and with the aid of a ruler the reader may care to rediscover them. With four aligned triplets on the same plate, the chance* of them being accidental falls to 10^{-6}. Taking account of the previously calculated 10^{-4} for the clustering near 1146+111, Hazard's 48″ Schmidt plate rules out the cosmological theory to within an uncertainty of about 1 part in 10^{10}.

All this came out of the handful of MC 2 quasars, the ones that arrived at the Institute of Astronomy at Cambridge in the odd circumstances in which we had to borrow £50,000 in order to finance the project. Did some fairy godmother wave her wand over those MC 2 quasars, loading them with information that could not have been distilled from any other handful? I find it hard to believe that this could be so. What was special about the MC 2 sample was that Cyril Hazard had them, and they were *all* that he had. Unlike so many others, Cyril was not choked with data. All his scientific possessions were stuffed, in some disorder I fear to the casual eye, into one small office.

If this be judged a reasonable point of view, then the fortuitous chance of 1 in 10^{10} needed to bolster the cosmological theory according to the Arp–Hazard data, must be repeated many times over, perhaps thirty times over, giving a chance of about 1 in 10^{300} of the cosmological theory being correct. So let us go no further. Belatedly, after fifteen years, let us give the cosmological theory a tardy burial and proceed to enjoy a few moments of rationality.

* There is a point of subtlety here. Each alignment discovered by Saslaw, if taken separately from the probability of the grouping at 1146+111 would be $\sim 10^{-2}$ as with the Arp–Hazard alignment. If one wishes also to include the probability of the grouping at 1146+111, however, the chance of each Saslaw alignment subject to an initial knowledge of the grouping is $\sim 10^{-1}$.

The Remarkable Things that Aberration can do

Quasars are evidently born in clusters as a consequence of events of extreme cosmic violence. In a former essay I described them as an hierarchy of white holes.* While the redshifts may have some component from the expansion of the universe, the main contribution to the redshift comes from the Doppler effect.

To avoid blueshifts, which are not observed, it is necessary to suppose that quasars emit light anisotropically, the emission being directed backward with respect to the motion causing the Doppler shift. A possible explanation for this behaviour might lie in a sweeping of intergalactic gas on the front side. If such sweeping were to produce anisotropic opacity, with the optical depth much greater at the front than the back, the main problem for the Doppler theory would be resolved.

Since 18 years of endeavour by many workers has failed to produce a satisfactory dynamical model of quasars within the framework of the cosmological theory, it would be unreasonable to expect a detailed model of the kind mentioned in the previous paragraph to be instantly apparent in the Doppler theory. What can be done immediately, however, is to resolve certain kinematical questions that at first sight seem puzzling.

It is commonly accepted that radiogalaxies expel oppositely-directed objects (or oppositely-directed beams of particles) from their nuclei. Some radiogalaxies have apparently experienced this

* F. Hoyle, *Steady-State Cosmology Revisited*, University College Cardiff Press, 1980.

phenomenon more than once, which shows that the nucleus can survive such events. Thus in any particular instance three objects are involved — the nucleus of the galaxy in question and the two objects expelled from it. And in the nature of the process the three objects are collinear. It follows that if in a cluster of rapidly-moving quasars one of them behaves like the nucleus of a radiogalaxy, expelling two oppositely-directed objects which are also quasars, the effect is to produce a triplet of aligned quasars.

On this interpretation, it has been A and Y, with the lower redshifts in Figure 4, that have played roles analogous to the nuclei of radiogalaxies. The same situation obtains for the two triplets to be found in Figure 3, suggesting that in a fragmentation cascade giving rise to a cluster of quasars the members with the highest speeds are those which have experienced the most fragmentations.

Provided quasars have lifetimes appreciably longer than a million years, groups like those of Figures 3 and 4 must be rare, since such groups fly apart quickly. Only those that are young can be observed in their early compact arrangements. As a group separates there are two special conditions, however, that permit particular group members to maintain a close association together on the sky. This will happen, for example, if two or more members have only small transverse motions with respect to the terrestrial observer. There is no constraint on the line of sight motions in such a situation, so that the redshifts would in general be quite different from one member to another, as in the quasar pairs of Table 1. Another possibility is for two quasars to arise from the gentle fission of a single object, in which case the two

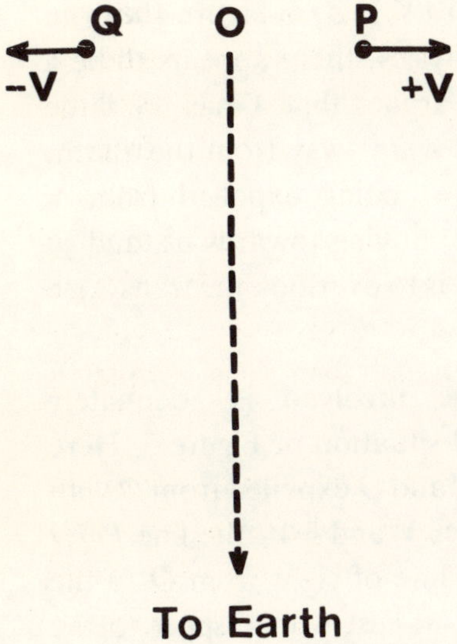

$$1 + z = \frac{1}{\sqrt{1 - v^2}}$$

Figure 5
In this picture the Earth is at rest, and the quasars *P* and *Q* are moving

quasars share an essentially common motion. In this second case the association is physical — the two quasars would show essentially the same redshift — whereas in the first case the association was only apparent.

The recent discovery by B. Margon, G.A. Chanan and R.A. Downes (Nature, 290, 1981, 480) of a quasar companion to the well-known quasar 3C 345 may be an example of the second case, as may be several other pairs to which Cyril Hazard has recently drawn my attention. The pair 0957 + 561 in Table 1 might conceivably be also of this type. While there are cogent spectroscopic reasons for thinking that 0957 + 561 appears double because of a gravitational lens, there are other aspects of this interpretation which cause difficulties, and perhaps 0957 + 561 arose from the gentle fission of a common parent into the two closely similar quasars of this pair.

If one looks at (A,B,C) or (X,Y,Z) in Figure 4 with an eye tuned to Newtonian physics, there appears to be a contradiction. Thus if one argues that X has the large redshift 2.1 because it is moving away from the Earth, one might think that Z — being expelled from Y oppositely to X — must be moving towards us, and so should be blue-shifted. This is to overlook the transverse Doppler effect, however.

The kinematic principle involved is adequately illustrated by the simplified situation of Figure 5. Here we have two equal objects P and Q expelled from O with oppositely-directed velocities V and $-V$, the line POQ being perpendicular to the line of sight from O to the Earth, with O being taken at rest with respect to the Earth. Neglecting the distance PQ in comparison with the distance from O to the Earth, light from both P and Q will appear to a terrestrial observer to have the same redshift z given by

$$1 + z = \frac{1}{\sqrt{(1 - V^2)}},$$

where V is measured in terms of the speed of light as unit.

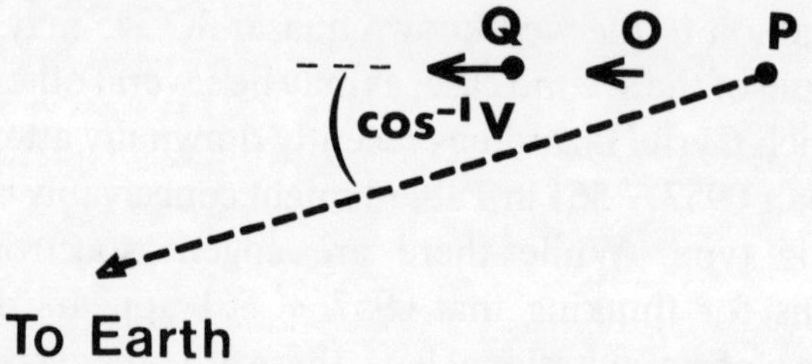

Figure 6
In this picture the quasar P is at rest. The midpoint O has speed $-V$ and the quasar Q has speed $-2V/(1 + V^2)$.

The question arises as to how Figure 5 is to be squared with the model described above. With the quasars emitting light in the fashion of the exhaust of a rocket, it might seem as if neither *P* nor *Q* could be seen from Earth. The effect of aberration now comes to the rescue, however. According to the theory of special relativity, an observer associated with *P* judges the light which reaches the Earth to begin its journey at an angle $\cos^{-1}V$ to the direction of the quasar exhaust, as illustrated in Figure 6. And like a miracle, an observer associated with *Q* has the picture shown in Figure 7, with the direction in which light must be propagated in order to reach the Earth swung around so as to make an angle $\cos^{-1}V$ with the rearward direction of *Q*!

As an example, put $V = 0.95$. Then $z = 2.203$ and $\cos^{-1}V = 18.19°$. Provided *P* and *Q* emit cones of light backwards, the cones having half-angles exceeding 18.19°, a terrestrial observer does indeed see these objects.

To develop this surprising situation more generally, Figure 8 is the terrestrial frame for a moving object *X*, while Figure 9 is the picture for an observer associated with *X*. The angle θ, at which *X* judges light must be emitted in order to reach the Earth, is related to α in Figure 8 by the equation of aberration

$$\cos\theta = \frac{V - \cos\alpha}{1 - V\cos\alpha} . \qquad (1)$$

And a terrestrial observer finds the light from *X* to have experienced a redshift *z* given by

$$1 + z = \frac{1 - V\cos\alpha}{\sqrt{1 - V^2}} . \qquad (2)$$

Keeping V fixed, imagine that α is decreased from π to 0. Initially we have $1 + z = \sqrt{[(1+V)/(1-V)]}$ and finally $1 + z = \sqrt{[(1-V)/(1+V)]}$, results that are well-known for the 1-dimensional Doppler shift. The final value requires z to be negative, i.e. a blueshift.

Evidently there is a particular value of α, α_0 say, giving $z = 0$. Thus putting $z = 0$ in (2), we have

$$V \cos \alpha_0 = 1 - \sqrt{1 - V^2}. \qquad (3)$$

The situation is that for α between π and α_0 the light from X appears redshifted, while from α_0 to 0 it appears blue-shifted.

If one puts $\alpha = \alpha_0$ in (1) a result emerges (after a certain amount of manipulation) which I did not find obvious. The value of θ corresponding to α_0, θ_0 say, is equal to α_0,

$$\theta_0 = \alpha_0. \qquad (4)$$

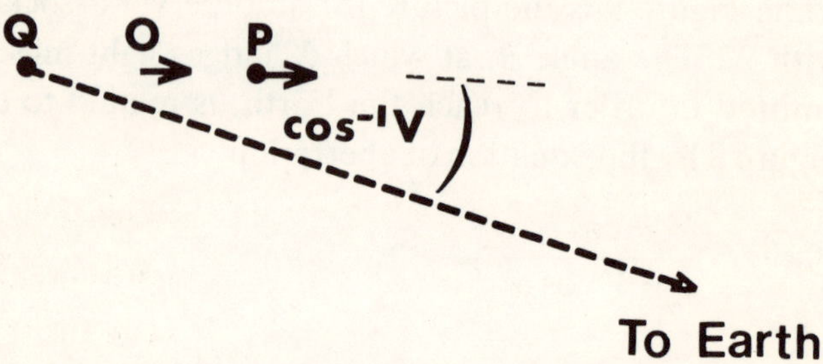

Figure 7
In this picture the quasar Q is at rest. The midpoint O has speed V and the quasar P has speed $2V/(1 + V^2)$.

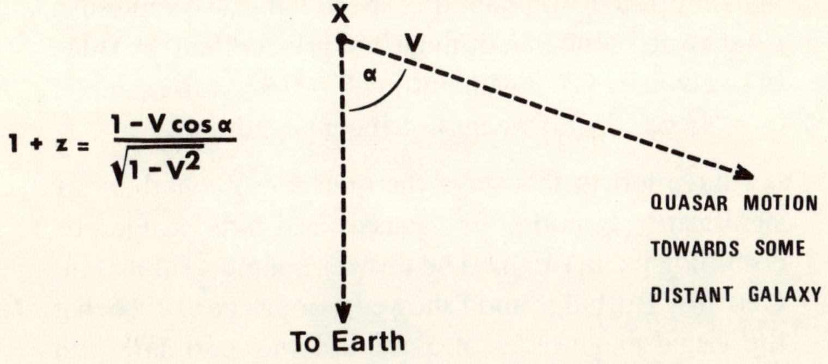

Figure 8
In this picture the Earth is at rest, and the quasar X has speed V in the direction defined by the angle α.

Figure 9
In this picture the quasar X is at rest. The angle α is the same as in Figure 8.

This result allows the requirement that no blueshifts be observed to be stated in an elegant and simple form. The light emitted from a quasar of speed V must be emitted in a backward cone of half-angle not greater than the value of α_0 given by (3). Again putting $V = 0.95$ as an example, $\alpha_0 = 43.62°$, by no means a stringent condition.

I have left to this stage the question of whether any significant fraction of quasar redshifts could be cosmological in origin. The answer is almost surely not. Geoffrey Burbidge and I showed as long ago as 1966 that the visual magnitudes of quasars do not correlate with their redshifts according to a cosmological relationship, and in 1972 I showed that the redshift-magnitude distribution of all quasars then known becomes a scatter diagram when a $(1+z)^{-2}$ factor is removed from the magnitudes. Because nothing different has emerged since then, and to keep the total number of quasars plotted within bounds, the result of 1972 is shown here in Figure 10.

The factor $(1+z)^{-2}$ is of course just the effect of motion in flat-space, and is what must of necessity be obtained in the present Doppler-shift model. No relation between z and the magnitudes remains after this effect has been taken out, proving that cosmological effects are negligible. In 1978, Wills and Lynds showed from the so-called V/V_m test that the number-magnitude distribution for quasars is closely consistent with their being in flat-space, pointing strongly to the same conclusion.*

* D. Wills and R. Lynds, *Astrophysical Journal Supplement Series*, 36 (1978), 317.

Two nearby quasars with different speeds of motion will appear to a terrestrial observer to separate unless the motions happen to be along the line of sight, from which it follows that searching for close pairs of quasars is equivalent to applying a strong selection for quasars moving directly away from us. According to the Doppler-shift theory, quasars in close pairs with different redshifts are therefore cases in which we are looking directly along the 'quasar exhaust'. Because their aspects are thus the same, we might then expect the members of a pair to appear more similar to each other than quasars do broadly one to another. This expectation is complicated, however, by the differing redshifts of the members, which cause us to see different portions of the quasar spectra, and which add the $(1+z)^{-2}$ factor to the bolometric magnitudes. Even so,

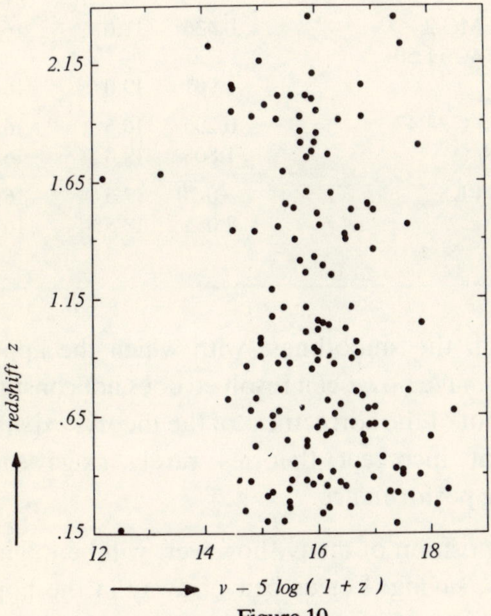

Figure 10

Redshift-magnitude relation for quasars

by taking the $(1+z)^{-2}$ factor out of the visual magnitudes we can make a rough test of this requirement. Table 2 shows the result. It can be seen that the magnitudes within each pair become remarkably alike, in agreement with the general theoretical requirement.

Table 2

Coord-inates	Name		Red-shift	Visual Magni-tudes	Visual Magnitudes after removing $(1+z)^{-2}$ factor
0254−334	(A)	PKS	1.915	17.0	14.7
	(B)		1.857	16.0	13.7
1038+528	OL	564	~0.68	17.5	16.4
			2.30	18.5	15.9
1318+290	TON	155	1.703	16.9	14.7
	TON	156	0.549	16.4	15.4
1548+114	(A)	MC 2	0.436	17.0	16.2
		4C 11.50			
	(B)		1.901	19.0	16.7
1721+342		4C 34.47	0.206	16.5	16.1
	No name		1.80	19.0	16.8
2143−156	(A)	PKS	~0.70	17.5	16.3
	(B)		2.055	18.5	16.1

Although the smoothness with which the apparent difficulties have now been resolved does not constitute a logical proof of the correctness of the theory, experience shows that incorrect theories rarely stage such a convincing performance.

The disposition of many, however, will be to gird up their loins, saying 'Let us beaver away in the hope of finding a disproof of this Doppler idea, so that we may then return in comfort of mind to the cosmological theory.' But no return to the cosmological theory is possible. This is just what the facts do not allow. If the Doppler theory fails for some proper reason, the gravitational theory (3), mentioned at the outset, must be reconsidered. If, in accordance with past experience, the gravitational theory also fails, then quasar redshifts must arise from a so-far unknown cause.

Appendix

Note on the Effect of Aberration on the Alignment of Quasar Triplets

Fred Hoyle and Jayant Narlikar

Let the quasar triplet (X, Y, Z) be generated through the ejection in opposite directions of the flanking quasars X and Z from the central quasar Y. The members of the triplet will appear collinear for all observers at rest with respect to Y, irrespective of whether Y received a recoil in the ejection process or not. An observer O, momentarily coincident with the Earth and at rest with respect to Y, will therefore see (X, Y, Z) collinear.

Suppose O moves with speed v towards a point P on the celestial sphere of an observer E attached to the actual Earth. Observer E does not in general see (X, Y, Z) collinear. Lines drawn by E from Y to X and Z respectively will have an angle between them that differs slightly from π, by an amount χ, say.

To investigate this problem, let both O and E set up a system of latitude and longitude on their respective celestial spheres, both using P as pole (which they can do since each observes the relative motion with speed v of the other). To O, the triplet (X, Y, Z) lies on a great circle, which O finds to intersect the equator, specified by P, at a point γ, say. Defining the zero meridian to be the great circle through P and γ, the situation for O is similar to the well-known arrangement of the ecliptic and celestial equator in elementary positional astronomy, with (X, Y, Z) having each quasar member lying on the 'ecliptic'. Write θ for the colatitude ($\pi/2$

minus 'declination') and ϕ for the latitude ('right ascension') of an arbitrary point on the celestial sphere of O.

The distribution of objects — stars, quasars etc — on the celestial sphere of O is mapped by aberration into a changed distribution on the celestial sphere of E. With the system of coordinates described above, this mapping involves a change of the colatitude, $\theta \to \lambda$ say, but no change of longitude, $\phi \to \phi$. The mapping $\theta \to \lambda$ is determined by any one of the following four ways of writing the transformation:

$$\cos\theta = \frac{\cos\lambda + v}{1 + v\cos\lambda}, \qquad \cos\lambda = \frac{\cos\theta - v}{1 - v\cos\theta}, \qquad \text{(A.1)}$$

$$\sin\theta = \frac{\sin\lambda\sqrt{1 - v^2}}{1 + v\cos\lambda}, \qquad \sin\lambda = \frac{\sin\theta\sqrt{1 - v^2}}{1 - v\cos\theta}.$$

The signs with which v appears in these formulae correspond to the motion of O relative to E being *towards P*.

Adjacent points (θ, φ) and $(\theta + d\theta, \varphi + d\varphi)$ on the celestial sphere of O map into points (λ, φ), $(\lambda + d\lambda, \varphi + d\varphi)$ on the celestial sphere of E, where from (A.1) it is easy to show that

$$\frac{d\theta}{\sin\theta} = \frac{d\lambda}{\sin\lambda}. \qquad \text{(A.2)}$$

Suppose this displacement is along the particular great circle on the celestial sphere of O that contains (X, Y, Z), and let κ be the angle which this great circle makes with the small circle of latitude at (θ, φ), so that

$$\frac{d\theta}{\sin\theta} = \tan\kappa \, d\varphi \ . \tag{A.3}$$

Because of (A.2) it is evident that observer E sees the map of the great circle as a curve which makes the same angle κ with the small circle of latitude at (λ, φ). Indeed the great circle seen by O and its map seen by E make the same angles at all corresponding small circles of latitude. At first sight one might think this result requires the map also to be a great circle, but this is not so!

It can be proved that the angle χ defined above (the amount by which the angle between lines drawn by E from Y to X and Z respectively differs from π) is given to first order by

$$\chi = \tfrac{1}{2} \quad \frac{v \sin\lambda}{1 + v\cos\lambda} \, . \cos\kappa . \ \Theta \, , \tag{A.4}$$

where Θ in the angular separation of X and Z, and κ in the angle with the circle of latitude at Y. All quantities in (A.4) are as seen by E. The sense of the deviation from collinearity is that Y lies on the same side of the great circle through X and Z as the pole P.

Suppose, for example, that $\lambda = \pi/2$, $\kappa = \pi/4$, $v = 1/\sqrt{2}$. Then $\chi = \Theta/4$. If one takes $\Theta = 12$ arc minutes as representative of the quasar triplets discovered by Arp and Hazard, we thus have $\chi = 3$ arc minutes for these particular values of λ, κ, v. The line from X to Y appears to E to miss Z by

$$\sim \frac{\pi}{180} \cdot \frac{\Theta}{4} \cdot \frac{\Theta}{120} \text{ arc minutes,}$$

or about 0.3 arc seconds.

Using (A.1) it is easy to show that (A.4) can be written in the form

$$\chi = \tfrac{1}{2} \qquad \frac{v}{\sqrt{1 - v^2}} \sin\theta \cos\kappa\,\Theta. \tag{A.5}$$

Although Θ must still be determined by E; θ, κ, and v in (A.5) can be determined by O. Because χ increases without limit as $v \to 1$, i.e. as v tends to the speed of light, the deviation from collinearity as seen by E is not subject to an upper limit. However, it would need a very exceptional situation for χ to be more than twice the value calculated in the preceding paragraph, which represents a reasonably typical situation. Thus the redshift of Y as seen by E, given by $1 + z = 1/\sqrt{1 - v^2}$, would be $z = \sqrt{2} - 1 = 0.414$, not greatly different from the redshifts of the central quasars in the triplets found by Arp and Hazard.

Although $\chi \simeq 0.5$ arc seconds represents a reasonable upper limit for $\Theta = 12$ arc minutes, the deviation χ increases with Θ. Indeed, the line from X to Y misses Z by an amount that increases as Θ^2. Thus for $\Theta = 24$ arc minutes the line from X to Y could miss Z by up to ~ 2 arc seconds.

The formulae (A.4) and (A.5) are so compact that one feels a short easy proof of their validity should be possible. We ourselves have been unable to find a simple proof, however. We have used independent methods, both of which have the property of highly involved systems of equations collapsing in the end into the simple form of (A.4) or of (A.5). Possibly the reader may have better luck at finding an easy route in this problem. The result (A.4) is a potential test of the theory described in the main text.